GUESS WHO?

At The Zoo

Jeannette Rowe

The bear growls.

Guess who growls?

Guess who is tall?

The giraffe is tall.

The elephant is BIG.

Guess who is big?

Guess who swims?

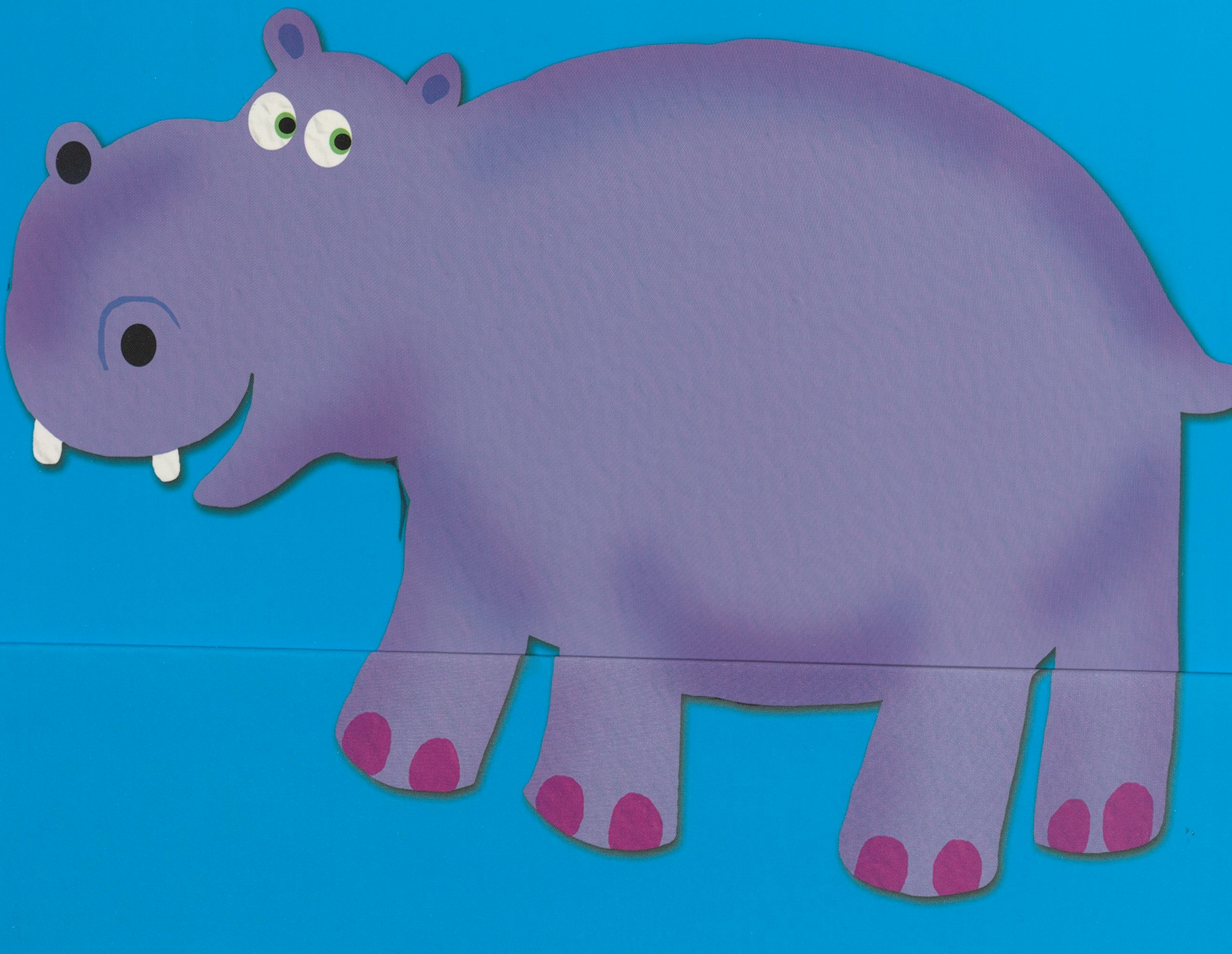

The hippo swims.

The crocodile smiles.

Guess who smiles?

Guess who chatters?

The monkey chatters.

The lion roars.

Guess who roars?

Guess who has feathers?

The peacock has feathers.

The zebra has stripes.

Guess who has stripes?

Guess who is cute?

The meerkats are cute.

The koala is fuzzy.

Guess who is fuzzy?

Guess who lives at the zoo?

GUESS WHO?

At The Zoo

Introduce your curious child to animals at the zoo with this delightful 'Flip-the-Flap' book by bestselling Australian preschool author Jeannette Rowe.

Jeannette ROWE

LOTHIAN Children's Books

This edition published in Australia and New Zealand in 2023 by Lothian Children's Books (an imprint of Hachette Australia Pty Limited)
Gadigal Country, Level 17, 207 Kent Street, Sydney, NSW 2000
www.hachettechildrens.com.au

First published in Australia in 2017 by Lake Press

Designed by Jeannette Rowe

Printed in China by Starlite Development

US: $14.99 CAN: $19.99

ISBN 978-0-7344-2260-6

9 780734 422606 >

www.hachettechildrens.com.au